Copyright © 2013 by All rights reserved. Illustrated by Book Writer Corner Published by Book Writer Corner info@bookwritercorner.com (878) 219-4793

A special shoutout to my incredible husband whose love brought this book to life! Adrion, so proud of the young man you have become! I'm so blessed to be your mom! Brian and Sofia, keep creating!

Mercedes Ochoa, Thank you for opening the door of STEM for me. To all the AMAZING STEM teachers from MUSD; thank you for inspiring me! Laura Toney, thank you for your guidance and support. Alyx Casillas, thank you for INSPIRING my new organization SKILLS!

STEM

What do you think a civil engineer does?
I've got the scoop! Civil engineers' secret weapon! Crafting amazing bridges! Let's go on a bridge tour today. Feast your eyes on this suspension bridge - it's all about those colossal foundations to kick things off. And don't forget, those steel connectors are the glue keeping everything solid up top!

How do you think this bridge was built?

The bridge started as wooden, then upgraded to iron, and finally evolved into a shiny steel masterpiece.

Check out the mighty girder bridge! It's like a giant made of beams, those super-strong steel or concrete rods. These beams are the unsung heroes, holding up the bridge and carrying the weight of all those zooming vehicles.

"Hey there, ladies and gents, we're cruising on an epic arch bridge!"

What in the world is this bridge crafted from? Time to put on your engineering goggles and spot the magic!

Civil engineers get crafty with stones, concrete, iron, or steel to construct a majestic arch bridge.

Hey there, young explorers! Let's rewind to our adventure yesterday. Remember what we discovered about those cool engineers? Why do you think civil engineers rock the world?

Let's ponder the epic items on our brainstorming table: cotton bags, straws, sticks, pipe cleaners and don't forget our trusty glue!

What's the grand plan for all this stuff? Got any burning questions about materials? Could we possibly construct a bridge out of it all?

Let's put our architect hats on and sketch out the masterplan for our epic bridge project! Picture this: our bridges will strut their stuff in a STEM showdown with other classrooms. It's a battle of creativity, and I've got my eye on the prize for our James Madison crew! So, grab your pencils, keep those creative juices flowing, and let's bridge the gap to victory with our out-of-this-world designs!

Once the detective work on the materials is wrapped up, it's showtime for the students at the classroom stage! Armed with their chosen materials, they have a tight 30-minute window to engineer their very own bridges. Let the building bonanza begin!

The bell chimed, bringing STEM class to an end. The students gleefully chatted about their bridge creations, eager to tweak them. Alas, the ringing bell meant they had to shelve their tinkering plans until next week.

Dr.Caos is beaming with pride for the fantastic job his students pulled off in his STEM class today.

Index of Terms:

Civil engineer
● Definition: A civil engineer is a professional who is dedicated to the design, construction and maintenance of infrastructure and civil works, such as bridges, roads, buildings and water and sanitation systems.

Arch Bridge
● Definition: An arch bridge is an engineering structure that uses a series of supporting arches to support the load of the bridge. Arch bridges are known for their strength and architectural beauty.

Beam Bridge
● Definition: A girder bridge is a type of bridge that uses horizontal beams to support the load. It is one of the most common bridge designs and is used in a variety of applications.

Suspension bridge
● Definition: A suspension bridge is a structure in which the bridge deck hangs from cables suspended from towers or pillars. Suspension bridges are known for their ability to span long distances.

Truss Bridge
● Definition: A truss bridge is a type of bridge that uses a cross-girder structure to support the bridge deck. This design is effective for heavy loads and is used in a variety of configurations.

Contact Information

author.maduenayoung.cecilia@gmail.com

www.ingramcontent.com/pod-product-compliance
Lightning Source LLC
Chambersburg PA
CBHW051837210526
45473CB00005B/1911